ELECTRICITY FOR KIDS
Facts, Photos and Fun
Children's Electricity Books Edition

SPEEDY PUBLISHING

Electricity plays an important role in everyday life

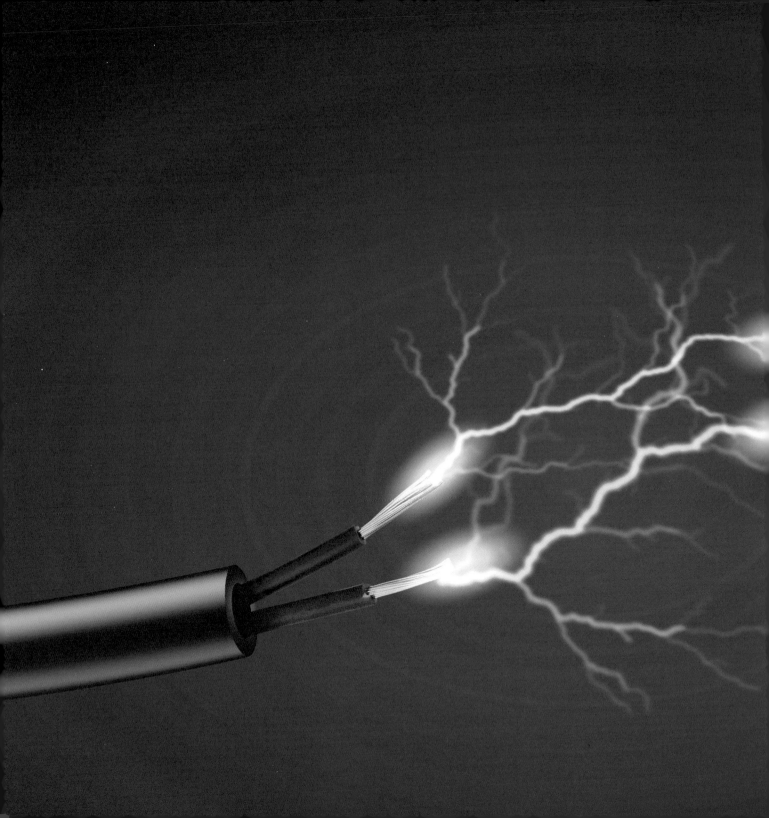

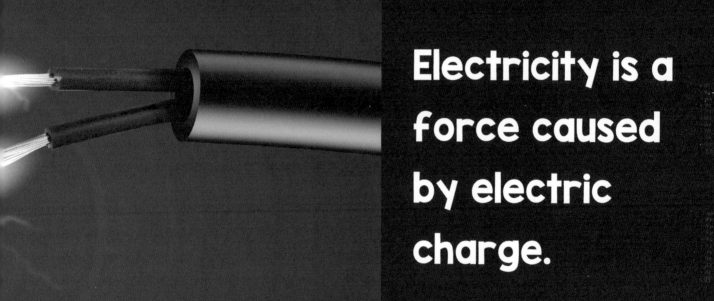

Electricity is a force caused by electric charge.

It is a form of energy which we use to power machines and electrical devices.

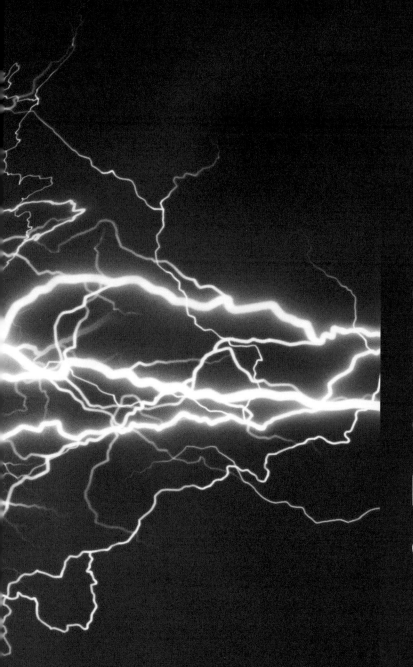

This energy is generated by the movement of positive and negative particles or electricity.

When the charges are not moving, electricity is called static electricity.

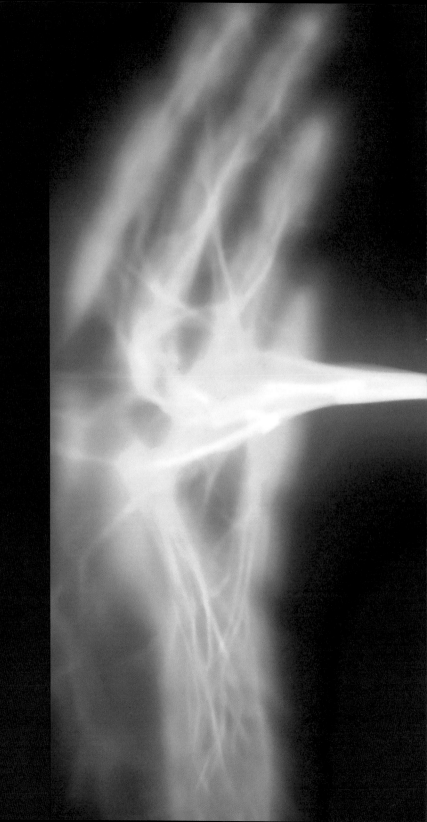

When charges
are moving
they are
an electric
current or
sometimes
called dynamic
electricity.

Lightning is an example of electrical energy found in nature.

Most forms of electrical energy in objects must be confined by wire when being used.

Electricity travels at the speed of light more than 186,000 miles per second.

Electricity can flow like water from one place to another, as a current in an electrical conductor.

Electricity
is mostly
generated
in places
called power
stations.

A spark
of static
electricity
can measure
up to three
thousand volts.

Our society relies heavily on the convenience and versatility of electricity. helps light your house, lets you watch TV and so much more.

The world's biggest source of energy for producing electricity comes from coal.

Solar energy is also used to create electricity. Solar cells convert light energy into electricity.

Wind turbines are used to generate electricity.

Around 80 different countries use wind power to generate electricity.

Hydroelectricity generates electricity by harnessing the gravitational force of falling water.

Most hydroelectric power stations use water held in dams to drive turbines and generators which turn mechanical energy into electrical energy.

Printed in Great Britain
by Amazon